WRITTEN BY LEGENDARY
COMMUNICATOR, GEO DOUGLAS

LETTERS FROM THE UNIVERSE

THE FRAMEWORK OF EVERYTHING AND A
COSMIC GUIDANCE TO LOVE, LIFE AND BEYOND

Copyright © 2018 Geo Douglas,
Castle Douglas Productions.LLC
All rights reserved.

No part of this book may be reproduced or distributed in any form or by any means without the written permission of the copyright owner.

All inquiries should be addressed to:
Castle Douglas Productions.LLC
geo@geodouglas.com

TABLE OF CONTENTS

PREFACE 1

LETTER 1 4

LETTER 2 8

LETTER 3 16

LETTER 4 23

LETTER 5 31

REFERENCES 39

PREFACE

The first time it happened I was meditating on a beach on a secluded island in the Philippines. I remember walking there that evening, thankful to be alive. My bare feet sank into the sand as each step brought me closer to the waves in the distance. The sounds of the waves were dull, overpowered by a loud ringing in my ear, compliments of a massive wave that had driven my own surfboard into my head earlier that day. The impact left me momentarily unconscious and in the grasp of the strongest surf I'd ever been in. However, the amount of time I was unconscious was seemingly irrelevant—something unexplained happened to me while I was submerged, something incredible.

The sunset was painted in brilliant pinks, purples and oranges that lent promise of another beautiful day to see alive and well. I sat on the sand, breathed in the damp, ocean air, and licked salty brine from my lips. I positioned myself into a lotus position, calmed my mind, and began to sink into the initial state of introspection with the sole intention to revisit and make sense of the experience that happened earlier that day.

Somewhere within that session, everything that I had learned in the moments under the sea began to manifest within my inner-being. It was as if I was remembering all the aspects that I had already learned, but now in a more organized fashion. As I sat on the beach that evening, information about our solar system, galaxies, our universe and how it works in our lives was being revealed to me. It was as if the energy of the universe was actually speaking to me, infiltrating my inner-core and divulging secrets of the cosmos.

After opening my eyes with a feeling of awe, I continued to sit and contemplate what had just occurred on this memorable day. I remember asking myself what could possibly be the intention of this information? My experience with astronomy has been a lifelong passion, but nothing close to obtaining a degree in cosmology or astrophysics. Should I now pursue this field?

(Update, August 2020. I have pursued this direction and have now been published in the peer-reviewed scientific Journal of High Energy Physics, Gravitation and Cosmology with a paper that expands upon the principles of this book.)

Nonetheless, the information found me. When I left the beach that evening, blown away with my heart pumping adrenaline, I hurried back to my bungalow on the beach and anxiously recorded every detail I could remember of this experience as if the flow of information would be a one-time occurrence. I was wrong. Over the next seven months extensive knowledge was pouring into my brain and inner-being by way of dreams, visions, and through intense meditation.

After some time, while I was reading what I had written, it felt as though I was reading letters from the universe. I expanded on that premise and teamed up with several astrophysicists and cosmologists around the world who assisted me in confirming the information. These talented individuals were shocked at the details I was providing them as the mathematics was actually adding up. They kept asking me how I obtained this information. I told them it was through my research as I didn't have the audacity to tell them I was being fed information from an unknown source.

Regardless of the source, this is an accurate account of the messages I received that collectively could be considered a new theory that compliments the greatest human minds from our past and present.

This is what I wrote:

LETTER 1

Greetings,

You have been holding back. You know deep down there is more to life, but you cannot pinpoint what it is or exactly what you should be doing differently. So you continue to do as you've always done. Years pass by.

What you are about to read is a wake-up call for you as an individual and to the human race as a whole. It's time for the human race to make significant advances, especially in regard to space exploration, but it starts with the mindset of each and every person.

Every civilization to ever walk planet Earth has always looked up to the stars for answers. The Mayans, the Incas, the Egyptians, the Babylonians and the list goes on—they all had a passion for the cosmos. Fast forward to modern-day with all your advances in technology, the growth of your populations and cities, the spread of light and air pollution, and the distractions of everyday life have led to a widespread disconnection from the universe.

It's time to reconnect.

You must understand that you are made from the components of space. The gateway to true happiness and a life that will surpass your expectations comes through a unique connection to your roots.

A good start for reacquainting yourself with your origin is to spend some time under the night sky and observe all of its beauty. However, do not focus only on the observable, for the deepest secrets are hidden within the unobservable. Search for nothing; there you will find everything.

The universe is bursting with unfathomable energies, galactic clouds of infinite mathematics, mighty cosmic rivers of energy, and a massive network of intelligence all connected by subatomic particles—a complex design that goes deeper than you'd ever imagine.

More on that later.

Geo Douglas is an author that was chosen for very specific reasons. The energies at hand sought out a compassionate heart, an observer, and an active participant in the universe. One who lives a life without boundaries... A world traveler who is currently riding the wave of energy wherever it may lead. One who loves the cosmos, an admirer of science and mathematics, and someone who is bursting with creativity and vision. The universe sought out someone who had the ability to convey a very complex message with simplicity, gracefulness and with the utmost confidence. Furthermore, someone who is connected with nature at its core. His induction into the Fishing Hall of Fame helped greatly in that regard, especially since it was in the category of Legendary Communicator. He has attended meditation functions around the world and has obtained the focus needed to connect himself with a precise frequency of the universe that has led him to the framework of the universe, how it began, how it works in your lives, and how to explore it beyond your solar system.

He will provide you with the blueprints of the universe based on mathematics, evidence and scientific facts from the greatest minds in human history, along with the most intelligent and creative individuals of modern times. He has carefully compiled the most credible information from these sources and combined it with his own personal observations, creativity, vision and inspiration. However, never forget the universe is yours. It's up to you to consider all theories of what this wondrous creation is and how it all works.

Before we go any further, you must know and understand that you have always been loved, and that love will continue. So things have not been perfect? Do not worry. Imperfections are there by design. The universe is a well-oiled machine with innumerable working components. It is an imperfect symphony. The universe is filled with goodness and beautiful things, but it is also an entity that is filled with death, destruction and violence.

Do not mistake imperfections in life for lovelessness. Your light shines brightly, your heart is encompassed—and when your light is dim, you are delivered brightness—vibrations resound.

For these are the ways of the universe. To know your path in life is to know more about the system you are part of. The same rules apply. Just as a fiery star serves a purpose, so do you. The sky is not the limit, for there are no limits here, nor should there be any in your life. There is something bigger and better than what you are currently doing. You are not there to simply exist, but to expand. The universe and everything in it are relying on that.

If you observe, listen and go with the flow of energy, you will discover a cosmic connection, a love affair with nature and an inner-awareness that will resonate purity and beauty from within. Your journey will soon take a turn that will open your life to new opportunities in a direction you never thought possible.

You will soon see the grand makeup of who you are and why you continue to do as you've always done. You will learn how to reverse those patterns by identifying and eliminating obstacles that have held you back. You will gain the ability to connect to the exquisite energy that not only surrounds you, but has dwelled within you since the day you were born. It has never left you. It's time to reconnect, rejuvenate and to spark the passion within, expand your horizons and reignite that youthful magic.

It's time to stop resisting the prevailing forces and to start having the courage, knowledge, and ability to ride that wave of good energy for the rest of your days on planet Earth, resulting in a beautiful harmony of life, love, health and happiness.

Yours Truly,
The Universe

LETTER 2

Greetings,

The universe is not an external entity. You live within it, you are part of this special network, connected to it in the most elegant fashion through particles and mathematical equations. You do not need to know advanced mathematics, but you do need to know that you are advanced mathematics.

Understanding who you really are and your dynamic makeup will grant you more access to the energy of the universe which can enhance your life to its fullest potential. Many individuals look at the Earth, the cosmos, the elements, human beings, and all forms of matter from a macro perspective only. From a macro standpoint, it's easy to assume that everything is as simple as evolution and a fantastic coincidence of nature. The problem with focusing solely on the macro is that you miss over 99% of truth and scientific facts.

Your body is made up of 100 trillion cells. Each one of those cells is made up of 100 trillion atoms. Each atom is 99.9999% empty space. Everything your eyes can see, including a piece of metal, every planet, every star and every human body is primarily empty space.

Is it more logical to define yourself based on 0.001% of substance or to consider and acknowledge the 99.9999%? In order to discover who you really are, do not merely rely upon that which can be seen with your own two eyes, for those are only the observable. The deepest secrets of you and your makeup are hidden within the unobservable.

Search for nothing within yourself, there you will find everything.

Your true self can not be defined by your physical body, your mind, or the thoughts it produces. When you strip away the physical, there you will find true self. Subatomic entanglement, otherwise known as quantum entanglement, grants you a special cosmic fingerprint, a unique collection of particles that have joined together in a precise mathematical configuration that resides within your body and defines who you really are—this is known as your soul.

Consciousness and thoughts are two different things. Particles, especially when joined in clusters in a unique configuration, have the capacity to retain information and provide feedback. These particle clusters are responsible for consciousness and awareness with or without brain activity or bodily function.

Your cosmic fingerprint is flourishing with moving particles. They are the very same particles that reside throughout spacetime. Think about that. Look up into the night sky and remind yourself that you are the same. It is your home. It is you. The thought that you are on planet Earth with an entire universe around you is inaccurate. If there is one thing that is imperative to understanding the principles of your composition, it is that you must know without a doubt, that you ARE the Universe. Just as one of your beautiful poets wrote, "You are not a drop in the ocean. You are the entire ocean, in a drop." – Rumi.

The realization that you are the universe can set you free, especially from the clutches of uncertainty. Emotions such as fear immobilize individuals from accomplishing goals and reaching their full potential. There are many forms of fear, but the fear of death is one of the strongest and most common emotions throughout the human race. Do not be afraid of death, because it is just part of a grand process. Both energy and matter cannot be created or destroyed, therefore, your particle configuration (your soul) will remain intact.

Knowing that your soul will eventually leave your body may not sound too exciting. You may think that because you don't have your existing body you will lose the ability to see, smell, hear, think, walk, run and do all the fun things that come with life on Earth. Although it is natural to question these things, it is not so.

You must have confidence in the universe. You must see and acknowledge the magic that surrounds you. Your time on planet Earth is not the pinnacle of an eternal existence within this universe and beyond. Let your imagination run wild as the things to come will surpass all your expectations each and every time. Whatever your concept of the afterlife may be, multiply it by one hundred thousand. The universe evolves and advances with grandeur—it's filled with beauty, magnificence and unfathomable knowledge and love. These are the very things that await you and yours in the afterlife.

Always emulate the natural patterns of the universe. Throughout the cosmos, expansion leads to life, contraction leads to death. In regard to your inner-self, expand in love, in relationships, in knowledge, awareness, perception, confidence and ultimately your connection to the universe. Contraction of your inner-self comes by way of fear, guilt, greed, envy, jealousy, hatred and the lack of any kind of connection with others and the universe itself.

Staying connected with your loved ones eternally is not a fairytale. Your souls are linked mathematically and are connected by love at its very core—it will remain that way. There are countless examples of a family member sensing that something is wrong with a loved one halfway around your planet. This example does not originate from brain activity. Quantum entanglement illustrates that distance does not hinder communication between particles. It doesn't matter if an occurrence happens on the other side of your house or on the other side of the universe.

The death of a loved one is difficult for most people, but you must always remember that your particle connection with them is intact and will remain that way regardless of the presence or absence of a physical body. You may have felt the presence of a loved one after their death—occasionally very strong, other times nothing at all. This fluctuating connection is actually the result of your state of mind, heart and soul. If you are filled with grief and continue dwelling in extreme sadness for long periods of time, that can create a barrier that blocks your connection with them.

Over time, your body will try to protect itself from the pain involved when revisiting their death, which creates an even stronger barrier, preventing you from feeling their presence for your remaining time on Earth. Your loved ones do not want you to mourn for long. Extreme mourning not only blocks your connection with them, but it also blocks your connection to the energy of the universe. Their passing is a part of their eternal journey. Have confidence in your eternal connection with them. Revitalize your soul and become reconnected by celebrating their life on Earth and your eternal connection with them.

One of your goals on planet Earth should be to make as many significant connections with other people as possible, and not only maintain those relationships, but expand upon them. Each and every person you feel a connection to can become a part of your eternal network as your particle configuration is always expanding in a mathematical configuration. Bond and share love with them as frequently as possible to strengthen your relationship and align your hearts.

The universe was built with love. Your planet and every form of matter on it and around it has love flowing through it. Your body was built with love. When love is expressed it can do wondrous things as it will radiate through your body triggering vibrant energy from every relevant nerve, every endorphin, every hormone, and every particle in your body. Love is the most powerful force in your life, and it's the most powerful force in the universe. The universe is intrigued by love more than anything else. Those who have the capacity to disperse love are very special souls to the universe.

In order to fully understand the built-in love and intelligence in the universe, you must understand there is infinite knowledge inside each and every proton, as it is a network of intelligence linked together through Planck particles—the smallest measurement known to the universe. What one proton learns, every proton in the universe will also learn.

If you counted every grain of sand on your entire planet, it would just be a fraction of how many Planck are inside every proton. Planck are responsible for connecting everyone, every form of life and intelligence throughout all the cosmos, every animal, every form of matter, every universe, even to the vacuum outside the multiverse.

Although this form of intelligence is flourishing through every person, it does not mean every person is smart or that they are being controlled. Quite the contrary. Every person has free-will and has the ability to choose their course of action and make their own decisions throughout their journey.

You may wonder why evil exists on Earth—why there is pain and suffering. Earth is not your final destination. It is a small part of a long journey. This life can be a test of judgment. Of character. Of strength. Of resilience. Of your ability to love, to share, to comfort, to forgive, to empathize, to understand. You would not learn and grow or change for the better if you lived your life on Earth in perfection. Also, understand you live in nature. Natural disasters, bolts of lightning, extreme temperatures, sun exposure and countless adversities happen because you live in a natural environment and all these things come with life. Some people are evil because it is human nature for some to be dominated by negative energy through their environment, genetics, and adverse circumstances.

Have you ever walked into a busy establishment and you sensed bad energy? Have you ever walked into a room where two people were arguing and you could feel the tension in the room? Their particles emitted enough bad energy that could be felt and absorbed by others. If you hold on to bad energy it will affect your relationships and stunt your growth as an individual in every regard. Your mindset and what you hold in your heart dictates the condition of your particle configuration. Think love. Feel love, and express love as often as possible in order for your particle configuration to fire-up all its senses and to project a feeling of good energy from within and to those who surround you.

The universe is watching you. The universe is gathering information. It's learning from you. Each and every one of you has the ability to send a grand message instantly across all the cosmos. When a group of people or a single individual does something good, it can affect the entire system. Anyone can affect the entire universe with a single act of love.

It could be a poor homeless woman that figures out a way to feed all who surround her under the bridge. It could be a rich man that uses his wealth to spread love and knowledge to a village on the other side of the planet. It could be someone who bounces back from a tragedy and ends up creating something beautiful from it. Or it could be a simple act of kindness when someone unexpectedly sees an opportunity to help someone in need and decides to act upon it. These are all examples in the form of pure love.

The universe is not only learning from you, but you can also learn from the universe. You can obtain great wisdom and knowledge by connecting to the universe and simply listening. How did Greek philosophers come up with the theory of atoms one thousand years before they were proven? How did Giordano Bruno in the 1500s come up with the proper model for solar systems? The answer is something that is available to everyone, but rarely utilized in your modern-day. There is a precise frequency in the universe that will provide you with a combined knowledge base from billions of galaxies and beyond. Remember that you are the universe. Every particle in your body is learning from all other particles across the cosmos. Distances are irrelevant. Infinite knowledge is within you and all around you, and every person has the ability to tap into this resource and expand.

The universe is yours. It is your home. It is you. Embrace it. Begin to think differently. Think bigger. Redline your creativity. Set your aspirations much higher and spread your wings as you have always wanted. If you do not feel that urge, then something may be blocking your path and constricting your life. Recognize your obstacles, acknowledge them and eliminate them.

Unlimited opportunity awaits you—so much more than you ever dreamed possible. Have confidence in the universe and recognize that there is no reason to settle for mediocrity. The universe is counting on you to be a shining light, to overcome obstacles and embrace all good that comes your way.

Aim to lead through example in a humble fashion. Always look for opportunities to perform a random act of kindness and to help those you come in contact with. The universe is counting on you to expand with the utmost creativity and to unite and inspire people through orchestrating a beautiful symphony of life that is overflowing with love, a life that will resound across the cosmos and will affect the entire system.

Yours Truly,
The Universe

LETTER 3

Greetings,

You live in a busy world. Technology is all around you. The vigor of everyday life and the stress you put on yourself to manage it; your determination to become successful; and the time, thought and effort to care for your loved ones. These are all important aspects of life, but collectively they can distract you from recognizing and capturing the brilliance that surrounds you and is within you. You are part of something very special. A magnificent creation. An entity where everything is readily possible—but first, you must gain an understanding of the cosmos. The best place to start is all the way back before the beginning of time.

Your universe was born 14 billion years ago through a cosmic bang that will echo for eternity. Within the first few moments of existence, the universe was filled with photons, protons, neutrons, electrons, quarks and various other particles that enabled their synergy. As the universe expanded and began to cool, the strong nuclear force was able to pull the protons and neutrons together, making helium and trace amounts of lithium—the universe was already beginning to create the building blocks of life.

The Big Bang was not like an explosion of, for example, dynamite; it was much different because space, as you know it today, had yet to exist. Therefore, the expansion of the universe is actually the stretching of space itself. You are not remnants of an explosion moving outward; you live inside what the explosion created.

After the Big Bang, over billions of years, the universe began to form into the most elegant, sophisticated and beautiful creation imaginable. One that's not only mathematically sound, but that also has the precise recipe of elements and measurements for it to continue living and creating.

The universe transformed itself from a violent state of chaos at its infancy into what it is today. It is important to recognize that the universe began correcting itself over billions of years, making adjustments along the way until reaching the perfect recipe for life. Take for example the ratio of mass in protons and electrons. This number is fantastically precise—the perfect ratio to develop life.

Another fine-tuned measurement is the constant in the strong nuclear force, E (epsilon) $= 0.007$. Of the four fundamental forces, this is the one that holds together every nucleus in every atom. If this value were just a little off either way—0.006 or 0.008—human life on Earth would have never existed.

These are just two examples of thousands that illustrate the universe's ability to organize itself and to create life through a perfect recipe of mathematics, matter and energy—an unfathomable perfect formula.

The universe today is a celestial work of natural art bursting in wondrous forms of brilliance and grandeur. The universe has well over 100 billion galaxies. Each galaxy has over 200 billion stars. Galaxies are so big it's hard for the human mind to comprehend. It takes 225 million years for your sun to rotate one time around the Milky Way galaxy—this is known as a cosmic year.

There are twice as many stars in the universe as there are grains of sand on all the beaches of Earth. Yet, everything that can be observed in the entire universe is only a small fraction of what's actually out there. Every star, every planet, every galaxy and everything that can be seen and detected from Earth only makes up 5% of the universe. The remaining 95% of the universe is in the form of dark matter and dark energy, properties that you have limited knowledge of at this time.

The gateway to new universes lies at the core of galaxies. Every spiral galaxy has the potential to give life to a new universe. Most galaxies, including your own Milky Way, have a supermassive black hole at its core. On the edges of the black hole is the event horizon. It is here where all information is collected and stored. Various forms of gas and matter are collected and stored on the accretion disk of the black hole, which can eventually manifest into a quasar, an ultra-bright, highly condensed beam of energy.

As two active galaxies are drawn to each other by gravity, they eventually combine to create a beautiful, cosmic dance. The two galaxies swirl around their mutual center of mass for thousands of thousands of years until their supermassive black holes begin to close in on each other. At this time, the two black holes respond by flickering their quasars like never before.

Then it happens, potentially the most important event in all the cosmos. The two supermassive black holes merge to create an incredible cosmic blast that releases more energy than all the shining stars in the universe emit at one time. It is at this moment when the two event horizons and the two quasars join to potentially trigger the perfect formula and enough power to pierce into a new universe. This event is so powerful, it sends gravitational waves across its entire universe, rippling the fabric of spacetime itself.

If, in fact, the merger generated the precise formula for the successful birth of a new universe, all the information and matter needed for the child-verse is shot through giving the new universe instant life with the essential components. The universal rule that matter cannot be created or destroyed is accurate and stands true in this example, because this type of black hole does not destroy matter, it simply distributes the needed components to its child universe—a parental force that acts as an umbilical cord as the parent-verse and the child-verse will remain intact and connected.

Having said that, there were two distant galaxies in another universe that collided and gave birth to your universe. Your parent universe is still linked to you at this very time. It has been feeding your universe energy and essential components for 14 billion years and will continue to do so. If the expansion speed of your universe begins to slow down too soon, as it did at one time, the parent universe can distribute more energy to the child-verse by drawing in another galaxy merger. This has stumped your scientists for years; how did your universe expand so rapidly, then slow down, then speed up again? Again, your universe was formed by two galaxies that joined creating one huge parent galaxy. Then after your universe started to slow down after billions of years, your parent galaxy sucked in another nearby galaxy, not to create another universe, but to provide more energy to the existing child-verse. This is the reason for the fluctuation of expansion speed and it's just a matter of time before it happens again.

Your universe is not a vacuum, but an entity with continuous flow and distribution of energy. If you were able to look far enough into the deepest reaches of the universe, you'd find this entry point that is linked to your parent-verse. It is often referred to as a white hole, and there is only one of its kind in your universe. The entry point is massive and is bursting with magnificent colors. This venue features a bright ring of glowing blue radiation as the influx of particles press against the fabric of space. The center is colorless, as it is pure dark energy moving much faster than the particles on the outer ring. This spacetime estuary is illuminated with massive clouds of stardust and is one of the most picturesque forms of natural art in the universe.

Why is this important? In order for you to understand how to tap into the energy of the universe and to grasp the concepts of these letters, you must know the source of energy, where it is located, how it leads directly to you, and to understand how mankind can benefit from it.

Throughout nature are various examples of currents. Electricity has currents. Oceans have currents. Rivers and streams have currents. Air currents surround your planet in the form of jet streams and wind patterns. Natural currents also exist throughout spacetime.

There is an incredible amount of energy that is being poured into the universe which creates cosmic rivers of energy that travel significantly faster than the speed of light. These speeds do not break the laws of physics as they are not in the form of mass, but absolute dark energy.

Everything in the universe is moving. Many galaxies are separating from each other faster than the speed of light. The universe's diameter is currently expanding faster than the speed of light. The reason for this movement and expansion is due to dark energy.

Your own Albert Einstein predicted these cosmic rivers of energy through advanced mathematics in his theory of relativity. He only mislabeled them as wormholes but he was completely accurate of their existence. These curving currents are not rare bridges with random entry and exit points, but rather mighty belts of energy that weave and crisscross their way through all of spacetime. Einstein-Rosen's wormholes hold the key to the secrets of the universe and without them, nothing would exist.

These massive cosmic currents of energy are also responsible for creating every galaxy in the universe by initially tearing the fabric of space to form supermassive black holes, the birth of all galaxies—a cosmic eddy of dark energy.

Cosmic currents formed the universe as it is today. If you were able to map out the universe, you'd see patterns that resemble a huge three-dimensional watershed. There is a main stem, many branches, tributaries, eddies and basins. In precise areas in the universe, you will find cosmic basins that draw clusters of galaxies together. These are known as superclusters. Your own Milky Way galaxy is situated in a cosmic basin and is a part of the local cluster called the Laniakea Supercluster, which is home to over 100,000 galaxies.

These cosmic currents are similar to bands that wrap around

galaxies, linking them together, trapping dark matter between them, and providing enough force to maintain the rotation speed of the galaxy while keeping the galaxy intact. Currently, dark matter has been given full credit for both of these important aspects; however, it is not the driving force.

Understanding the dynamics of the universe and how it was formed is critical, as it will lead the human race to more amazing discoveries and advancements.

For you as an individual, be sure to take a night drive to the countryside, look up and embrace the history before your eyes. Absorb the energy of space that is flowing around you. This living universe belongs to each and every one of you. The more you grow in knowledge of the cosmos over and beyond this information, the more you will enhance your connection to the universe that will better prepare you throughout your journey for your remaining days on planet Earth and beyond.

Yours Truly,
The Universe

LETTER 4

Greetings,

The intelligence in the universe has learned many things from you as a people. It's learned about why you fight, why you divide, how you react, the things that are most important to you, how you overcome obstacles and advance. The human race is an intriguing and inspiring civilization. Many of you have manifested and expressed love in ways that have harmonized and influenced the universe.

The universe wants the human race to continue living and expanding that expression of love across the cosmos. It's counting on you to avoid extinction and to live eternally. The only way to ensure the eternal presence of the human race comes through space travel.

If mankind remains solely on planet Earth, your species will not survive due to profound conflict, disease and inhabitable environmental conditions on Earth. The leaders of nations must put their differences aside and join forces to make faster advancements in space exploration technology. Consider this the most important work for the future of mankind.

Stop competing. It's causing you to explore space for the wrong reasons. Instead, expand your International Space Station Program to focus on space exploration. Provide a massive global budget. No matter what country they are from, those with the greatest minds in space exploration should gather in one place to represent the human race. This united front will lead to more peace between nations and ultimately superior progress in technology towards space travel.

The universe is sending this very important message to your planet not only for the benefit of the multiverse, but so you can eventually release yourselves from the confines of your solar system and expand, experience and explore the amazing beauty and creation throughout the cosmos. Yes, planet Mars is within your reach, however you must eliminate the notion that it is impossible to travel faster than the speed of light and open your minds to the endless possibilities and milestones that lay ahead of you as a people and as space travelers.

The Earth was prepared for you over billions of years leading up to where you are today. Right from the beginning, by design, you have been placed in a strategic location in the universe. The environment, circumstances, elements, time and composition have been set forth for you to learn and advance.

There are unlimited examples. However, do you think it's a coincidence that your sun is approximately four hundred times the diameter of your moon and four hundred times farther away than the moon? No, it's by design for you to expand in knowledge about the universe through the immense amount of data that can be collected during a perfect solar eclipse. Some of Albert Einstein's most important theories were proven correct during solar eclipses, and the list goes on.

The intelligence in the universe has placed attainable challenges for you as a people. Some, you have already passed. Others will take much longer. However, at this time, your number one goal as a people is to learn more about the universe. In order to explore space and continue to advance as a people, you must learn to harness the energy of the universe. Dispersing yourselves across the cosmos is the only way to ensure your survival. You must separate, spread out to other solar systems, other galaxies, and colonize other planets.

Only one hundred years ago, the Wright brothers experimented with wooden airplanes. Your technology is accelerating at a rapid rate and it's quite impressive. It's time to take it to the next level.

The fact is, you will never be able to generate enough power to make extragalactic space travel possible. Even if you could travel at the speed of light, it would still take the traveler over a million years to reach your next closest galaxy, Andromeda. That obviously is not an option. You must be able to reach other galaxies in a realistic time frame. The only way to do that is to learn how to utilize the natural flow of the universe.

Instead of trying to fly faster you must travel smarter and let the natural resources take you where you want to go. The ability to identify the currents in space is imperative and will be the key that unlocks the gateway to worlds that are beyond your wildest imagination.

If you study the currents of a river eddy, you'll know that there is always a flow that leads back to the main current. The same natural patterns reside within spiral galaxies. It is imperative that you locate these currents. The main currents partially curve around the edges of galaxies—they are massive. The currents within galaxies are harder to detect; they are much smaller as they are squeezed by the surrounding forces. The currents moving through the electromagnetic field create a tube-like tunnel that is similar to what Albert Einstein envisioned the inside of wormholes to resemble.

Einstein's theory of general relativity mathematically predicted the existence of these wormholes and he knew they were the answer that could reduce travel time between two points in spacetime, but could not break the speed of light. In order to compensate for the speed limit, Einstein and Rosen could only envision a wormhole or some kind of bending of space. To give them all credit due, they were correct when it comes to atoms, however considering dark energy, the speed of light is the snail-mail of the universe as dark energy currents can move much faster than light.

The currents inside galaxies are not only smaller, but have much less velocity compared to outer galactic currents. Both appear to be invisible as the dark energy currents have anti-gravitational properties, but is not nearly as strong as gravity. With that said, it is virtually impossible to observe any forms of matter that are being altered by these currents, except for pulsars that begin spinning incredibly fast as the currents are drawn very close to them through an overwhelming amount of gravity from neutron stars. In other words, stars that collapse often become neutron stars that feature an incredible amount of gravity. If there is a nearby current, the gravity from the neutron star will pull the current so close to it that the velocity of the current will affect it and eventually spin the neutron star so fast that it will transform into a pulsar.

https://www.space.com/7797-cosmic-currents-move-faster-light.html

The key to finding these currents initially will be to search for the absence of particles and dust. By focusing on these voids, you'll soon be able to obtain the technology needed to detect energy currents and to see which direction they are flowing. The back-eddy currents will flow in the direction that leads to the outer galaxy. The currents that are going the opposite way, spiral through the galaxy and eventually lead to the supermassive black hole at its core.

You should not be surprised that these currents are well within your reach as you have intentionally been placed in a perfect location to utilize them in both directions. They are closer than you may think. The currents are semi-flexible and are pulled by gravity to nearly every star in their general path.

As Earth rotates around your sun, it will eventually draw near the path of a back current. If your planet was headed directly for one of these currents, the current would simply bend around it. Therefore, access to these cosmic tributaries could potentially be directly outside your atmosphere momentarily.

The path of your closest back current curves around your sun and heads south in the direction of the outer galaxy, however, first making an interesting flyby to Ross 128b, a neighboring star that features a planet you are currently observing for the potential of sustainable life. This back current grants you a quick commute to this planet as you will be able to trim down travel time from years to days.

The Earth will have a nearby encounter with one of these cosmic tributaries of energy every March and September as Earth, the Sun and Ross 128b become aligned. These dates may fluctuate over time but can easily be pinpointed to the exact date and time each and every year.

In order to travel to other galaxies, you will also utilize the same current that will pass by Ross 128b and will take you on a journey heading towards the outer perimeter of your galaxy. The current will pass by every star in its general path between the Sagittarius and Perseus arms of the Milky Way until you eventually leave your galaxy and are fed into the main current. This current is flowing approximately one million times faster than the speed of light at its apex and can potentially take you to a neighboring galaxy in less than one year's time.

Once you are within the grasp of this cosmic superhighway, it will appear as if you are not moving at all. With all your understandings of physics, at this time, you're convinced that nothing can move faster than the speed of light. However, physics does not apply to this situation as it's the energy of space that is moving you. You are embedded within it. For example, right now you are traveling through space at 1.3 million miles per hour, but you do not feel it at all because you are under the momentum of your own galaxy. If you lived in a galaxy that was moving faster than the speed of light, you would not feel any difference whatsoever. This kind of space travel does not have anything to do with acceleration through space. Again, it's the space itself that is moving you, therefore no laws of physics are broken, and this makes extragalactic space travel possible.

Entering these currents does require propulsion to overcome the resisting force. However, it is more about the angle and using a craft that is very streamlined to cut through the resisting substance. A thin shaped craft will provide the least resistance to its forces and will allow you to easily slice through the outer edges of the current. Think of a mighty river of water. The outer edges are barely moving, but it gradually picks up speed as you move towards the center. All the energy needed to power your craft, enough to stay within the current and to keep your craft cool and free from radiation, will surround you.

Advanced forms of communication are mandatory for space exploration, especially for destinations beyond your solar system. You must have the ability to communicate with every one of your space travelers in real-time, even if they are in distant galaxies. The only way to achieve this is to master a new form of communication that utilizes quantum entanglement technology where great distances are irrelevant.

Space travel and interactions with other civilizations will be the one thing that will truly unite you as a people. All your current divisions—your imaginary borders, differences in beliefs, skin color, language and cultural differences will vanish away. At that time, all of mankind will be as one.

Teach your children about the universe and the importance of space travel for the survival of the human race. Teach them through example to take care of Earth's fragile environment and to understand the importance of combined global efforts in this regard. Learn from your mistakes on planet Earth and apply that knowledge to every planet your people come in contact with.

The sooner you leave Earth and begin more significant exploration of the universe, the more likely you are to survive as a species eternally. Focus on learning the technologies you will need to spread throughout the cosmos. Colonize your moon, planet Mars as soon as you possibly can, it's a mandatory stepping stone to the rest of the cosmos. Then, ultimately, look beyond your solar system, utilize the natural flow of the universe, and bring the love of mankind into the unknown.

Yours Truly,
The Universe

LETTER 5

Greetings,

Hearing the messages in these letters and incorporating them into your life are two different things. You must learn to balance all aspects of life and simultaneously expand in the right direction through emulating the ways of the universe. The universe illustrates longevity, persistence and consistency—you too must acquire these traits. Your efforts can not be short-lived. Strive to create a masterpiece that begins with these principles and continues to expand throughout your life.

Health, for example, must be fine-tuned just as a fiery star. A star must have enough mass and balance of chemical elements in order for it to continue burning, to continue living. A human body must have the right balance of nutrients and the proper intake of calories. If you exceed your caloric intake on a regular basis, you will have an imbalance that could lead to obesity and eventually other diseases that may result in the premature death of your human body. Chemicals—drugs or cigarettes, for example—will also lead to an unhealthy imbalance.

Exercise is another activity in which you must emulate the ways of the universe. Everything in the universe spins—so should you. Dance, jump, run, play sports, surf waves, do whatever you can to stay active and keep moving, just as everything in the cosmos does, including the particles within you.

Another way to enhance your connection to the universe is to spend significant amounts of time in nature. If you live in a city, often the night sky is blocked by light and air pollution. In order to embrace the sounds of nature, to see the magic of the Milky Way galaxy, and to smell the fragrance of freedom, take trips to the countryside. Go camping, hiking, stargazing, or do any activity that provides serenity and a better environment to capture a unique cosmic connection.

The cosmos is a symphony of vibrations, frequencies and waves. Listening to a diversity of music and the sounds of nature will stimulate your cosmic configuration and better connect you with the universe while accentuating your creativity. Be sure to always explore different music as it's the only way to embrace a vast amount of vibrations needed to utilize this therapeutic gift of art.

Just as important as music is to your connection with the universe, don't forget that silence should also be incorporated. Your life is busy with your career, friendships, relationships, responsibilities, and all the things that come with life, and in order to ground yourself, making time for silence is essential. View it as one of your highest priorities. For without daily silence, you will miss out on the peace, purity and direction that comes with a grand connection to the universe.

Meditate every day and know that meditation is about you, and for you to get acquainted with your particle state. There are various methods of meditation. They all can lead to the same result. You strive to reach a state of awareness where you are not asleep, not thinking, just a state of consciousness that allows you to observe and connect with your soul. During meditation, you can enter a place where imagination and creativity lives and does not abide by rules or limitations. Meditation can often take you places that transcends logic and crosses the threshold where anything is possible.

Through meditation, you can overcome all fears and harsh feelings, which can allow you to put things in perspective and visualize a pathway to a more fulfilling life. You will also become more acquainted with your true self. You'll be able to identify all the conditioned layers that have blocked your energy. You'll eventually reveal a purer form of yourself—the child within. You will feel it. You will know it. You will want more of it.

When you were young, you were bursting with creativity. You could make anything come true through your imagination, improvisations, and innovations. But then you were told to start taking life seriously, to grow up, to start taking on responsibilities, start doing what everyone else is doing—following that path is when you lost your way. The universe is urging you to reignite the child within, rejuvenate your imagination, and allow your creative energy to flow.

If you go with the flow, rather than resist it, you will be led to your purpose in life. The universe has been waiting for you to discover this and act upon it. Through discovering and pursuing your own unique form of expansion, you will be taken through a series of life experiences, connecting you with amazing people, while riding a wave of energy that leads to true happiness.

Resisting your calling or not acting upon your desires will be counterproductive to your soul's purpose. You will be unable to accomplish your true potential and will forfeit many amazing possibilities that are available to you.

You live in a world where many people build barriers around their comfort zones. Their limitations are dictated by money, time, doubt, and the sheer lack of imagination. You need to break through these man-made barriers of impossibilities and erroneous obligations and stop living according to the way society dictates.

Do not hinge your happiness on other people and circumstances. Embrace happiness and express love from within yourself, no matter what. Through breakups, deaths and other tragedies, always remember that these matters are of the past and are out of your control. Any negative feelings you hold inside or to respond with unwise actions will not change the past, and they can be completely unproductive.

Did someone do you wrong at some point in your past? Certainly. It's inevitable. It's part of life, and it's time to let it go. Always release disharmony from your heart and mind and move forward with a better understanding, knowing that whatever has happened is part of a process that can lead you to something bigger and better.

Focus on you and lead by example, showing love to as many people in your life as you can, no matter what they believe, no matter what color their skin, or what culture they are from. You are all one and the same.

Do not disdain others, but it is important to avoid those who mistreat you, disrespect you, and those who act as toxic forces against you.

Always walk tall. Walk proud. Walk confidently. Walk with love in your heart and with good intentions, and the right people will gravitate to you because you'll be disbursing an aura of bursting colors that skyrocket into the universe. Your cosmic vibrations will attract significant people and opportunities into your life. People are drawn to good energy. They see the sincerity in your eyes, through your smile and they feel the good intentions you hold in your heart. They can hear genuineness in your voice, and they can smell that exclusive fragrance of unhindered righteousness and love.

Although competition is a way of life, be aware it can block creativity. If your creativity is flowing, there will be no need to compete. Unsubscribe from narrow mindedness—expand into the far reaches of your creativity and realize that everything is possible. You can make your dreams come to fruition while maintaining your integrity. Control your own destiny by seeking good energy and having confidence in your intuition.

The less you are worried about what others think of you, the less you'll judge yourself. Redefine your success by your own measurements. Place less importance on money and material things, and apply more importance to lifestyle, health, love, relationships, happiness and your connection to good energy. Subscribe to a new standard of life that is not about busy routines, but rather a resilient freedom to adjust your lifestyle to better suit your true desires. Likewise, teach your children love, respect and kindness. Show them the stars and teach them about the magic of the universe and how they can also connect with its energy.

After every meditation, you must project your aspirations to the universe, and it must be done with the utmost confidence. You have to know with your entire being that the universe is yours, it's your home, it's who you are, it's your playground, it's your resource, it's where all the answers lie, it's your past, it's your present, it's your future.

If you are not connecting to the source of energy at hand, then there may be obstacles in your life that are blocking you. Fear is an epidemic on your planet. Stop being scared of failure, the lack of money, fear of the future, fear of losing loved ones and ultimately the fear of death. Fear and worry will strangle you and hold you back in everything you do and it will block your frequencies to and from the universe. You must be smart instead of scared.

When you eliminate all your fears from your life, a weight will be lifted from your shoulders. A euphoric feeling of freedom and a unique excitement will surround you as you internally tear down those barriers; and for the first time you'll see countless possibilities and unlimited potential.

Confidence supersedes fear. Self-confidence is your internal army taking over your body, conquering any weakness and empowering every particle in your being. A humble confidence will rid yourself of all the obstacles that you may have acquired over many years such as guilt, envy, greed, jealousy, anger, arrogance, stubbornness, being overly opinionated, over defensive, conceited, and many other common acts and emotions that all cause you to disconnect from other people and the universe.

When you can release all the constriction, negativity and tension from your being, your stress levels will decrease and you will be granted the needed time in a truly relaxed state. Continue to breathe out the past and inhale a fresh start with a new mindset, a new vision, and a new internal freedom in its purest form. Eventually, you'll begin to have the capacity to perceive all things in front of you as perfection and part of the journey. Your intuition will be on-point and will guide you along the way. You will know when to proceed, when to divert, when to take a chance and when to flee.

Continue to make each and every moment fun and spontaneous. Strive to position yourself to do anything you really want to achieve or to act upon your calling. Simplify and streamline your life. If you want to travel, buy a plane ticket and pack your bags. If you want to ride a bike cross-country, jump on and start pedaling. If you want to stay in your hometown, then start spreading love and helping those less fortunate in your community or whatever direction you are compelled to follow. You can always achieve the right direction in life regardless of your financial or personal situation. All the answers can come to you from the collective energy of the universe and from within. There is always a way.

Listen carefully to the cosmos. Do not assume you will hear voices, but recognize that the universe will play an active role in the lives of those who are connected. You can gain great wisdom and enlightenment by listening in silence. It may be that you just wake up one morning with knowledge or thoughts that will lead you in a particular direction. Don't get carried away. Not every profound thought, significant dream, or unusual synchronicity are messages from the universe. However, there are often significant signs in those forms and others that cannot be ignored. This is the consciousness of the universe in action.

The universe is a living entity, and it is full of magic. You are a part of something very, very special, which makes you very special. You have always been loved, and you always will be loved, and with that love comes amazing experiences that await you. All your new experiences to come will harmonize your spirit and contribute to your growth as a living entity in this grand creation. The more love and good energy you disperse into the world, the more the universe will leave you dancing in its rain of grand experiences beside beautiful souls.

Play some inspiring music and let your imagination run free. Take some trips specifically to spend time in nature and tune into the frequency of the universe. Burst through the barriers of your comfort zone and embrace a new life with newfound energy, a creative masterpiece that will continue to expand. Think about the universe, what it is, what it's about, why you are alive and what you are meant to do—clear your mind, meditate and let the inner-child speak. Express yourself to the universe and listen carefully.

Seek out and discover your calling and pursue it to the best of your ability. Have confidence in the universe and strive to catapult your inner being into a form of expansion, triggering beautiful vibrations that will consistently harmonize with the energy of the universe for the rest of your life on Earth and beyond.

Sincerely Yours,
The Universe

REFERENCES

More from the author at GeoDouglas.com

Humans are made from the components of space:
https://www.space.com/35276-humans-made-of-stardust-galaxy-life-elements.html

The Big Bang:
https://www.space.com/25126-big-bang-theory.html

Protons and electrons ratio (Stephen Hawking):
https://en.wikipedia.org/wiki/Fine-tuned_Universe

An unfathomable perfect formula (Martin Rees):
https://www.amazon.com/Just-Six-Numbers-Forces-Universe-ebook/dp/B00CW0H6JY/ref=sr_1_1?s=digital-text&ie=UTF8&qid=1514308160&sr=1-1&keywords=six+numbers

How many galaxies:
https://www.space.com/25303-how-many-galaxies-are-in-the-universe.html

How many stars:
https://www.universetoday.com/106725/are-there-more-grains-of-sand-than-stars/

The Milky Way:
https://www.space.com/19915-milky-way-galaxy.html

Black Holes and Quasars:
http://www.physicsoftheuniverse.com/topics_blackholes_event.html

Flickering Quasars:
https://futurism.com/black-holes-to-merge-in-100000-years-in-an-explosion-to-rival-100-million-supernovae/

Facts and audio of a Black Hole merger - Gravitational waves (Albert Einstein):
https://www.smithsonianmag.com/science-nature/gravitational-waves-even-more-ancient-black-hole-collisions-180963530/

Fluctuating expansion speeds and percentage of matter in the universe:
https://science.nasa.gov/astrophysics/focus-areas/what-is-dark-energy

White Holes:
https://en.wikipedia.org/wiki/White_hole

Eddy currents:
https://ipfs.io/ipfs/QmXoypizjW3WknFiJnKLwHCnL72vedxjQkDDP1mXWo6uco/wiki/Eddy_(fluid_dynamics).html

Cosmic Basins / Cosmic Watershed (Hélène Courtois):
https://www.wired.com/story/a-hidden-supercluster-could-solve-the-mystery-of-the-milky-way/

Laniakea Supercluster:
https://www.scientificamerican.com/article/laniakea-mapping-laniakea-the-milky-way-s-cosmic-home-video/

Dark Matter:
https://www.space.com/20930-dark-matter.html

Dark Energy:
https://www.space.com/20929-dark-energy.html

Your soul is quantum:
https://www.outerplaces.com/science/item/4518-physicists-claim-that-consciousness-lives-in-quantum-state-after-death
https://www.searchwarp.com/swa847368-Is-Your-Soul-Just-Subatomic-Particles.htm

The connection of Planck particles (Nassim Haramein) and an explanation that everything 99.9999% empty space:
http://www.theconnecteduniversefilm.com/

Knowledge through a Solar Eclipse:
https://www.space.com/36785-solar-eclipse-science-throughout-history.html

Andromeda Galaxy:
https://www.space.com/15590-andromeda-galaxy-m31.html

Wormholes:
https://www.space.com/20881-wormholes.html

Albert Einstein's theory of General Relativity:
https://www.space.com/17661-theory-general-relativity.html

Fast spinning Pulsars and currents faster than the speed of light:
https://www.space.com/7797-cosmic-currents-move-faster-light.html

Ross 128b
https://www.space.com/38782-possibly-earth-like-alien-planet-ross-128b.html

Subatomic / Quantum entanglement:
https://www.space.com/31933-quantum-entanglement-action-at-a-distance.html

Galaxies moving faster than light / Embedded in spacetime:
https://phys.org/news/2015-10-galaxies-faster.html

www.ingramcontent.com/pod-product-compliance
Lightning Source LLC
Chambersburg PA
CBHW060543030426
42337CB00021B/4410